T0413363

THE POWER OF THE SUN

The Sun and Earth's Surface

Jodyanne Benson

New York

Published in 2020 by Cavendish Square Publishing, LLC
243 5th Avenue, Suite 136, New York, NY 10016
Copyright © 2020 by Cavendish Square Publishing, LLC
First Edition
No part of this publication may be reproduced, stored in a retrieval system, or transmitted in any form or by any means—electronic, mechanical, photocopying, recording, or otherwise—without the prior permission of the copyright owner. Request for permission should be addressed to Permissions, Cavendish Square Publishing, 243 5th Avenue, Suite 136, New York, NY 10016. Tel (877) 980-4450; fax (877) 980-4454.
Website: cavendishsq.com
This publication represents the opinions and views of the author based on his or her personal experience, knowledge, and research. The information in this book serves as a general guide only. The author and publisher have used their best efforts in preparing this book and disclaim liability rising directly or indirectly from the use and application of this book.
All websites were available and accurate when this book was sent to press.

Library of Congress Cataloging-in-Publication Data

Names: Benson, Jodyanne, author.
Title: The sun and earth's surface / Jodyanne Benson.
Description: New York : Cavendish Square, 2020. | Series: The power of the sun |
Audience: Grades 2-5. | Includes bibliographical references and index.
Identifiers: LCCN 2018044269 (print) | LCCN 2018047414 (ebook) |
ISBN 9781502646699 (ebook) | ISBN 9781502646682 (library bound) |
ISBN 9781502646668 (pbk.) | ISBN 9781502646675 (6 pack)
Subjects: LCSH: Sun--Juvenile literature. | Earth (Planet)--Juvenile literature. |
Earth (Planet)--Rotation--Juvenile literature. | Solar energy--Juvenile literature.
Classification: LCC QB521.5 (ebook) | LCC QB521.5 .B48 2020 (print) |
DDC 523.7/2--dc23
LC record available at https://lccn.loc.gov/2018044269

Editorial Director: David McNamara
Editor: Lauren Miller
Copy Editor: Nathan Heidelberger
Associate Art Director: Alan Sliwinski
Designer: Jessica Nevins
Production Coordinator: Karol Szymczuk
Photo Research: J8 Media

The photographs in this book are used by permission and through the courtesy of: Cover PhilipYb Studio's/Shutterstock.com; p.3 (used throughout) Black Prometheus/Shutterstock.com; p. 4 NASA/JPL-Caltech/ESA/CXC/STScI; p. 6 Vadim Sadovski/Shutterstock.com; p. 7 NN/Wikimedia Commons/File:Helios-Metope, Troja, Athena-Tempel.jpg/Public Domain, (used throughout) St-n1ce/Shutterstock.com; p. 8 Vjanez/iStockphoto.com; p. 10 Peter Hermes Furian/Shutterstock.com; p. 11 Planetary Visions Ltd/Science Source; p. 13 Marques/Shutterstock.com; p. 14 Wapcaplet in Blender/Wikimedia Commons/File:Earth tilt sample.jpg/CC BY SA 3.0; p. 15 John Dommers/Science Source; p. 16 Stephen J Krasemann/Getty Images; p. 19 Bakhtiar Zein/Shutterstock.com; p. 20 Simply Recorded/iStockphoto.com; p. 21 Jamesdvdsn/ iStockphoto.com; p. 23 Patty Chan/Shutterstock.com; p. 24 Vadim Sadovski/Shutterstock.com; p. 26 (left to right) Natalia Sokko/iStockphoto.com, Denis Burdin/Shutterstock.com; p. 28 Aoldman/iStockphoto.com.

Printed in the United States of America

Contents

Our solar system is located in the Milky Way galaxy.

Chapter 1

The Sun and Our Planet

BILLIONS of stars are found in the Milky Way galaxy. Even more are sprinkled throughout the universe. But one star is very important to our planet. That star is called the sun. The sun is the closest star to Earth. This very hot ball of gases is at the center of our solar system.

With the help of the sun, Earth has everything needed for life. The sun helps crops grow. It also provides energy and warmth. The sun even causes weather patterns and seasons.

Energy from the sun heats Earth's surface.

Earth's Surface

Earth's surface, called the crust, is the outer layer of our planet. It's the part that we see every day. Land and water make up Earth's surface. The land is made up of rock and soil. Landforms like mountains and valleys are created by movements of Earth's crust. They can also be formed by elements like gravity, wind, and water.

Water covers about 70 percent of Earth's surface. Oceans and seas are mostly salt water. Freshwater is found in rivers, streams, ponds, lakes, and **glaciers**. About 2 percent of Earth's water is fresh. Only 1 percent is drinkable!

Let's find out more about how the sun interacts with Earth's surface.

SUN POWER

It takes about eight minutes for light from the sun to reach Earth.

Helios: God of the Sun

Humans have always been fascinated by the solar system. Early civilizations focused on the sun because it's our source of light and heat. Everything on Earth needs the sun to survive. Many cultures developed **mythologies** about the sun. In Greek mythology, Helios was the god of the sun. Ancient Greeks believed he rode a golden chariot that brought the sun across the sky each day from the east to the west. In Greek and Roman art, Helios is pictured as a young man wearing a crown of sunrays.

Helios is shown in Greek and Roman art wearing a crown of sunrays.

SUN POWER

Many early cultures had a sun god. Ancient Egyptians called their sun god Ra.

We experience day and night because Earth rotates on its axis.

Chapter 2

Day and Night Cycles

HAVE you ever wondered why we see the sun in the sky during the day but not at night? Earth is a sphere that rotates on an imaginary line called an **axis**. This imaginary line passes through Earth's North and South Poles. These points are the most northern and most southern parts of the planet. As Earth rotates, different places are exposed to the sun. When one place rotates toward the sun, it becomes daytime. Then, as Earth keeps rotating, the sun sets there and it becomes nighttime.

The time of day is different depending on where you are on Earth. When one side is experiencing daytime, the other side is experiencing nighttime.

Earth's Rotation

Earth is constantly spinning. One day on Earth is twenty-four hours. This is how long it takes Earth to rotate once around its axis. Earth rotates at a speed of about 1,037 miles

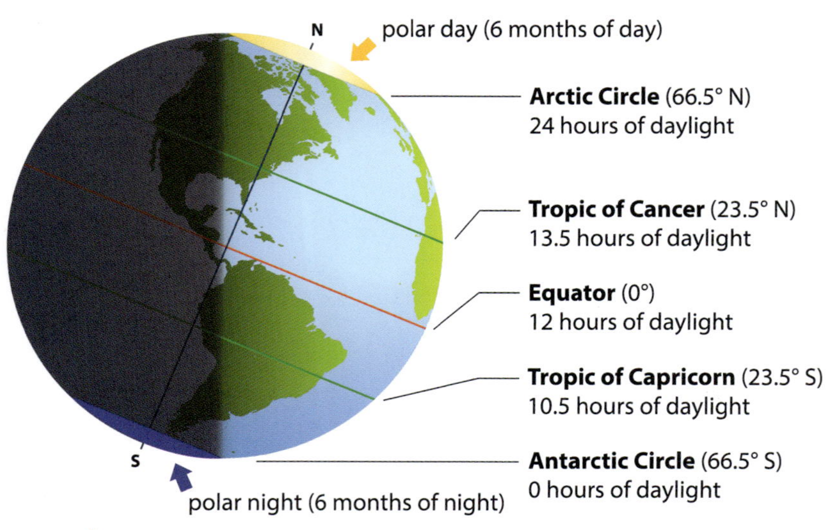

On the summer solstice, usually June 21, Earth's North Pole is tilted toward the sun. This means the Northern Hemisphere experiences the most daylight hours.

(1,669 kilometers) per hour at the **equator**. We can't feel Earth spin. Why? It's like riding in a car. Because the car is always moving, our bodies get used to the speed. Inside the car, we feel like we aren't moving. On Earth, we are moving, but we don't feel it because we are used to the speed.

The terminator line is curved because Earth's atmosphere bends sunlight.

An imaginary line separates the side of Earth experiencing daytime and the side that is experiencing nighttime. This line is called the **terminator**. It is also called the

SUN POWER

The surface of the sun is about 9,932 degrees Fahrenheit (5,500 degrees Celsius).

"gray line" or the "twilight zone." This line isn't perfectly straight. Earth's atmosphere slightly bends sunlight. This means that more than half of Earth's surface is always touched by sunlight.

The Sun's Effect on Earth's Surface

Have you ever walked barefoot on a sandy beach on a hot day? Did you want to run into the water to cool off? You might be wondering why some surfaces on Earth are hotter than others. Let's find out!

Light from the sun is called **solar energy**. Earth's surface warms as it receives more energy from the sun. Once the sun goes down, Earth's surface cools because it is no longer receiving energy from the sun. Earth's surface is coldest around sunrise. It is warmest in the late afternoon because more heat has been absorbed.

Earth's surfaces absorb and reflect the sun's energy in different ways. For example, water needs a lot of energy to heat up. So, the temperature of large bodies of water, like oceans, does not change very much between night and day. Sand, however, needs less

SUN POWER

Though most people think of the sun as yellow, sunlight actually contains all colors mixed together.

Dry material like sand in the desert absorbs the sun's energy quickly.

The Midnight Sun

There are some places in the world that experience twenty-four hours of sunlight! This phenomenon is called the midnight sun. It happens during the summer in parts of Alaska, Russia, Denmark, Norway, Sweden, Finland, Iceland, and Canada. In cities such as Tromsø, Norway, the sun never dips below the horizon from late May to July.

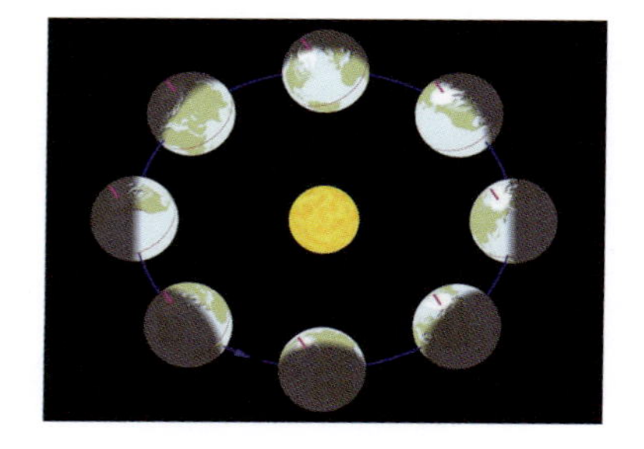

The model Earth at the top shows the position of the Northern Hemisphere in summer.

The midnight sun occurs because of Earth's tilt. Earth's axis is tilted 23.5 degrees. This is why globes are tilted to the side. The North Pole is pointed toward the sun around June 21, so the sun never sets. The South Pole, on the opposite side of the world, is pointed away from the sun. There, the sun does not appear at all during this time.

energy to heat up. This is why sandy deserts like the Sahara in northern Africa are very hot. Sand also cools very quickly. This makes these deserts cool at night.

ACTIVITY: Trace Your Shadow

Materials

- Different colors of sidewalk chalk
- A safe concrete area to trace shadows

Instructions

Grab a friend to learn about how shadows change at different times of the day. Outside on the blacktop, trace each other's shadows in the morning, at noon, and at the end of the school day. Make sure to mark your foot placement so you stand in the same place each time. Use the same color chalk to write the time next to the shadow. Talk to your friend about how the shadows change at different times of the day. Where was the sun at each time of day? Try to guess what will happen to your shadow in the evening.

Minnesota has cold, snowy winters because it is so far north.

Chapter 3

Seasons and Earth's Surface

EARTH has four seasons: winter, spring, summer, and fall. We wouldn't have these seasons without the sun. They follow expected patterns of change that happen year after year. These changes affect an area's temperature and **precipitation**. They can be predicted. For example, it is common during winter for Minnesota to get lots of snow.

Earth's Tilt and Orbit

About every 365 days, Earth completes a journey around the sun, called an **orbit**. One full trip, or orbit, is called a revolution. During the journey, Earth experiences the seasons. Seasons are caused by the tilt of Earth's axis (23.5 degrees). During Earth's orbit, the amount of sun hitting Earth's surface changes. Throughout the year, Earth's axis points toward or away from the sun. When the North Pole tilts toward the sun, the part of Earth north of the equator, called the Northern Hemisphere, experiences summer, and the Southern Hemisphere, the part of Earth below the equator, experiences winter. This is because the sun shines more directly on

SUN POWER

In ancient times, it was believed that the sun rotated around Earth instead of Earth rotating around the sun.

the Northern Hemisphere than the Southern Hemisphere. When the sun shines more directly on the Southern Hemisphere, the Northern Hemisphere experiences winter.

During its revolution, Earth is always roughly 93 million miles (150 million km) away from

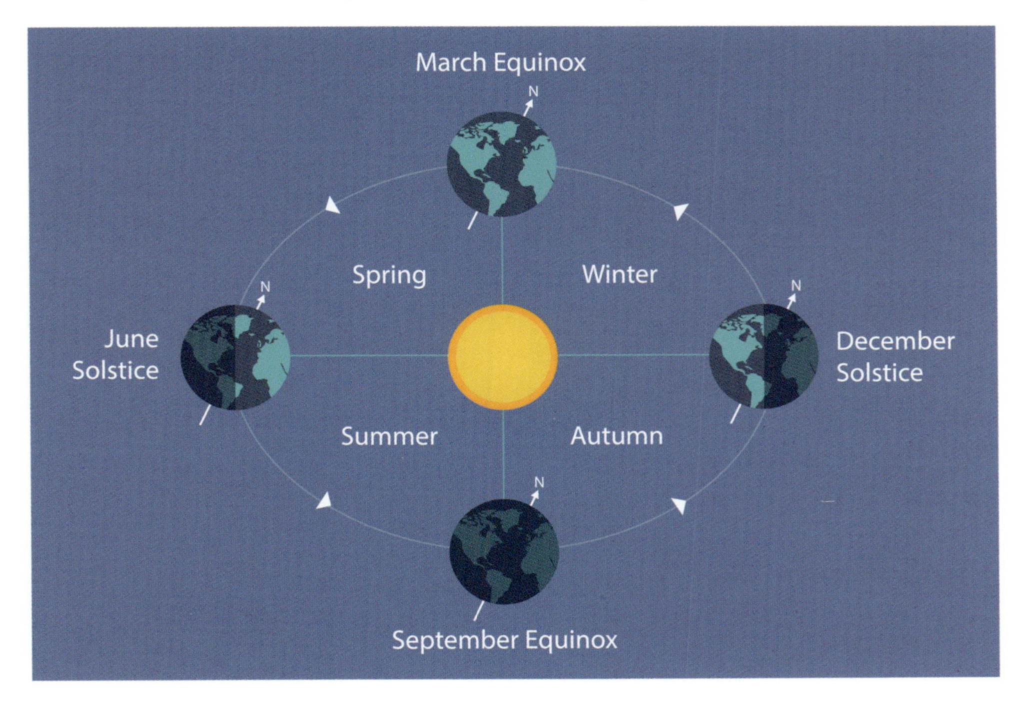

During the spring and fall equinoxes, there is an almost equal amount of daylight and darkness on Earth. The solstices mark the first days of summer and winter.

Change in Earth's Surface

Earth's surface is constantly changing. Even Earth's climate has changed throughout our planet's history. Evidence has shown that the Sahara, the world's largest hot desert, was once green and tropical. Scientists believe that Earth's position in relation to the sun was a reason for this change. About every one hundred thousand years, Earth's orbit changes back and forth from being circular to being more elliptical, or egg-shaped. The angle of Earth's axis also varies about every forty-one thousand years between 22.1 degrees and 24.5 degrees. These different positions once made it easier for plants to grow in the Sahara.

The Sahara has become hot and dry as Earth's orbit and tilt have changed over time.

the sun. So, the amount of energy that Earth receives is pretty constant.

The Angle of Sunlight

The angle of the sun's light affects how much energy a hemisphere receives. (Remember that light is energy!) Direct sunlight is warmer than sunlight striking Earth at an angle. In

Death Valley in California gets extremely hot because the rock surface radiates heat. The valley then traps the heat.

SUN POWER

The sun will eventually become a big star called a red giant. A change in temperature causes red giants to appear more red and orange.

the Northern Hemisphere, the noonday sun is higher in June than in December. During summer, the sun's rays strike the ground more directly. This causes Earth's surface to become warmer.

Some places on Earth have very extreme temperatures because of how the sun heats the surface. Death Valley in California had the highest temperature ever recorded in North America. It reached 134°F (56.7°C) on July 10, 1913. Heat from the sun radiates from the rock and gets trapped in the valley.

ACTIVITY: Flashlight Fun

Materials

- Flashlight
- Piece of paper

Instructions

Take a flashlight and piece of paper and shine the light directly on the paper. The light is bright, right? Now tilt the paper so that the light is more spread out. What differences do you notice?

The sun is a giant ball of hot gases.

Chapter 4

Solar Energy and Fuel for Earth's Surface

ALL year round, the sun provides energy and warmth to Earth.

Absorbing and Reflecting Sunlight

Fifty percent of the sun's energy is absorbed by the land and oceans that cover Earth's surface. The atmosphere and clouds absorb 20 percent of this energy. Thirty percent gets reflected back into space by Earth's surface, clouds, and the atmosphere.

The amount of energy that is absorbed or reflected depends on color and texture. Darker objects absorb more energy than lighter objects. Shiny objects reflect more energy than dull or rough objects. These differences cause changes in temperature, weather, and climate. For example, land and oceans absorb more energy than they reflect. Snow, ice, and clouds reflect more energy than they absorb. The sea ice in the Arctic absorbs less solar energy. So, the surface there

Oceans absorb more sun energy than they reflect. Snow and ice reflect more energy, making the surface cooler.

stays cooler. Blacktop absorbs more energy than it reflects. Blacktop can get so hot that you could burn your feet walking on it barefoot!

SUN POWER
The sun will eventually die, but it will be millions of years before that happens!

Earth's land surfaces have changed from human activity and natural processes. Things like air pollution, cutting down trees, and volcanic eruptions impact how much of the sun's energy is absorbed or reflected. For example, after an eruption, volcanic ash prevents the sun's rays from reaching Earth's surface. This causes temperatures to drop in the area around the volcano.

When you wake up in the morning, think about how your side of the planet is rotating into view of the sun. Or when you go to

sleep at night, remember that your side of the planet is rotating away from the sun. And when the leaves start falling and it becomes colder where you live, think about Earth's axis and how it is tilting away from the sun. The sun is an important part of our daily life. It is the reason we live happily on Earth's surface.

Without warmth from the sun, grass and trees would not be able to grow in our parks.

SUN POWER

Did you know that there are explosions on the sun's surface? These are called solar flares.

Glossary

axis The name for the imaginary line on which Earth rotates.

equator An imaginary line drawn around Earth, equally distant from both poles, that divides Earth into the Northern and Southern Hemispheres.

glaciers Slowly moving masses or rivers of ice.

mythologies Stories about history or religious figures like gods and goddesses.

orbit The path Earth follows around the sun.

precipitation The amount of rain, snow, and other water that falls on an area.

solar energy Power from the sun.

terminator The dividing line between the part of Earth receiving sunlight and the part that is not.

Find Out More

Books

Branley, Dr. Franklyn M. *Sunshine Makes the Seasons*. New York: HarperCollins Publishers, 2016.

Demuth, Patricia Brennan. *The Sun: Our Amazing Star*. New York: Grosset & Dunlap, 2016.

Websites

National Geographic Kids: Sun

https://kids.nationalgeographic.com/explore/space/sun/#sun.jpg

PBS Learning Media: Why Do We Have Seasons?

https://www.pbslearningmedia.org/resource/npls13.sci.ess.seasons/why-seasons/?#.W4mzyuhKhPY

Index

Page numbers in **boldface** refer to images. Entries in **boldface** are glossary terms.

About the Author

Jodyanne Benson is a writer and editor from Wisconsin. A longtime teacher, Benson believes that great books spark wonder, curiosity, and learning. She enjoys reading and spending time with her family.